PACIFIC COAST BIRD FINDER

Identifying Common Birds Along the Pacific Coast

ROGER J. LEDERER

illustrated by JACQUELYN GIUFFRÉ and CAROL E. BURR

HOW MUCH CAN A POCKET-SIZE BOOK DO?

This book will help you to identify 63 of the most common species in this area.

To identify *all* of the 500 or so bird species found on the Pacific Coast, you'll need a larger, more comprehensive field guide such as:

The Sibley Field Guide to Birds of Western North America, by David Allen Sibley. Alfred A Knopf, 2016.

For general information about Pacific Coast birds, we suggest:

Birds of California Field Guide, by Stan Tekiela. Adventure Publications, 2022.

Birds of Oregon Field Guide, by Stan Tekiela. Adventure Publications, 2022.

Birds of Washington Field Guide, by Stan Tekiela. Adventure Publications, 2022.

For general information about birds of North America, we recommend:

National Audubon Society Birds of North America (National Audubon Society Complete Guides), 2021.

Birds of North America (DK North American Bird Guides with the American Museum of Natural History), 2020.

 • ISBN 978-0-912550-35-0 • Printed in China • LCCN 2024006441 • naturestudy.com

HOW TO USE THIS BOOK

Like comprehensive field guides, *Pacific Coast Bird Finder* is organized in taxonomic order, beginning with grebes and ending with songbirds. There is no one way to identify birds, so when first using this book, you will have to page through it to identify the bird you are viewing. Once familiar with the book's organization, you will be able to go to the waterfowl or hawk descriptions quickly.

Because there are about 500 species of birds in the Pacific Coast area, bird-watching could become overwhelming. But you have to start somewhere, and a simple book like this is a good choice.

Each page has a sketch of the bird and gives its common and scientific names; body length from beak to tip of tail (L) and wingspread (W); some identifying features such as eye stripes, wing bars, tail pattern, and behavior; and icons that indicate its usual habitat.

Because the birds in this book are among the most common you will see while bird-watching, they should be easy to find and identify. As you develop your skills as a bird-watcher you will be able to identify other birds and move on to a comprehensive field guide.

Note: Pacific Coast means mainly the US coast but some of Canada and Mexico as well.

Pied-billed Grebe

Podilymbus podiceps

Riding low in the water, the small chunky grebe is identified by its brownish body and a white bill with a black patch, hence the pied name. Feeds on invertebrates, frogs, and fish; eats its own feathers to protect its stomach from bones. Builds a floating nest of reeds in emergent aquatic vegetation. Babies ride on parents' backs after hatching.

L: 14" W: 22"

Freshwater Marshes

Grebes are excellent swimmers and divers but rarely fly, except to migrate. Swimming aided by lobed toes and legs positioned far back on the body. Able to adjust buoyancy by trapping water between the body feathers to float and compressing feathers to sink.

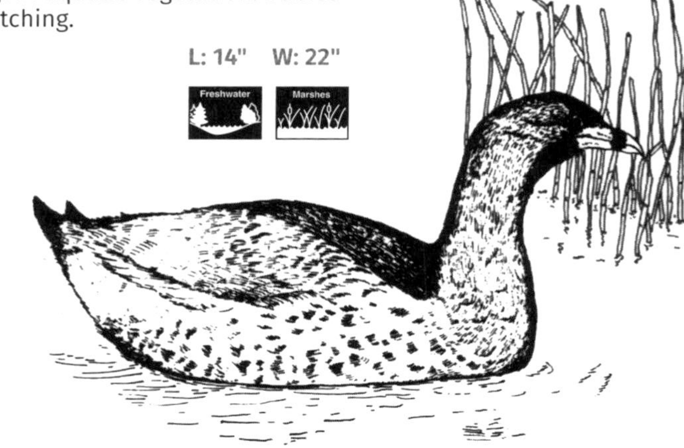

Western and Clark's Grebes

Aechmophorus occidentalis and *A. clarkia*

These long-necked grebes winter along the entire US Pacific Coast but breed eastward to Minnesota on inland lakes. Black on the back from the crown to the tail, with white undersides. The two species are virtually identical, except the black on the head of the Western Grebe extends below the eyes; the black is above the eyes on Clark's Grebe.

While courting, breeding pairs (or two competing males) swim side by side, making similar head movements. Often the birds will rise up and run across the surface of the water together until they simultaneously dive.

L: 25" W: 23"

L: 51"   W: 79"

Brown and White Pelicans

Pelecanus occidentalis and *P. erythrorhynchos*

The Brown Pelican inhabits the seashore, while the White Pelican lives inland and in some coastal areas. The Brown Pelican has a white head and neck and brown body; the all-white pelican has black wingtips. Both fish in groups and nest on rocky shores. The Brown Pelican feeds by diving for fish from the air and trapping prey in its expanded scoop-like pouch. White Pelicans sit on the water, often encircling their prey; stick their heads under the surface; and eat 4 pounds of fish a day. Before DDT was banned, the pesticide caused eggshells to thin so the parents crushed them while incubating.

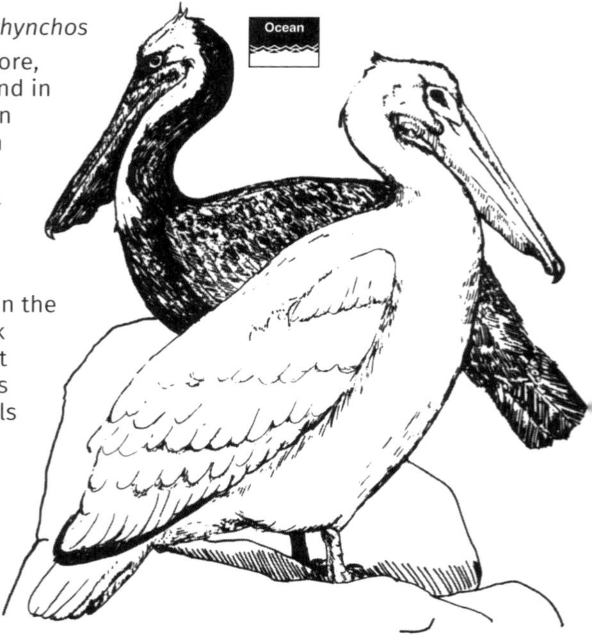

L: 63"   W: 108"

Double-crested Cormorant

Nannopterum auritum

Black with an orange face and throat; feather crests not obvious. Common along the coast and inland in the winter, they fish in groups and fly like pelicans in a straight-line flock. Often stand on rocks to dry their widespread wings. Like pelicans, cormorants have a throat pouch, although smaller. Cormorants dive for prey and either swallow it whole or grasp it in their hooked beaks, kill it, and then swallow it. Wide mouths help them scoop and swallow fish, and webbed feet help them swim rapidly. Similar **Brandt's Cormorant** (*Urile penicillatus*; L: 34", W: 48") has a blue throat, while the smaller **Pelagic Cormorant** (*Urile pelagicus*; L: 28", W: 39") has a red face and a white flank patch.

L: 33" W: 52"

EGRETS

Great Egret

Ardea alba

Three white egrets are common over most of the US. The Great Egret has black legs and feet and a yellow bill; the **Snowy Egret** (*Egretta thula*; L: 24", W: 41") has a black bill and legs with yellow feet. Almost exterminated in the early 1900s for their plumes to decorate hats and clothing; this "feather trade" led to the formation of the Audubon Society. The **Cattle Egret** (*Bulbulcus ibis*; L: 20", W: 36") has yellow legs and bill and, in breeding plumage, yellow plumes on the head and back. Native to Africa, Cattle Egrets have spread across the US since their introduction in 1941. Like herons, egrets silently stalk their aquatic prey in shallow water.

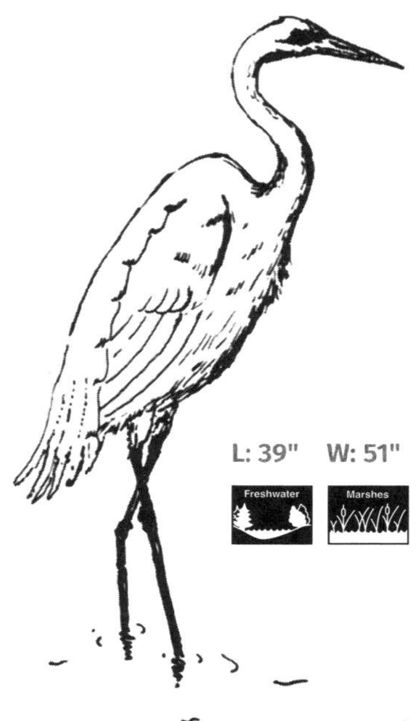

Great Blue Heron

Ardea herodias

Widespread throughout North America, the Great Blue Heron is all gray with a white face and dark crown that trails short plumes. Both herons and egrets stalk fish, crayfish, snakes, and other animals near or in shallow water, and they nest in large colonies. Great Blue Herons quietly stalk their prey with slow, silent footsteps, then plunge their bill into the water to capture it. Although it's a large bird at nearly 4 feet tall, its hollow bones allow it to tip the scale at only 5–6 pounds.

There is no biological difference in herons and egrets; the names just come from different languages.

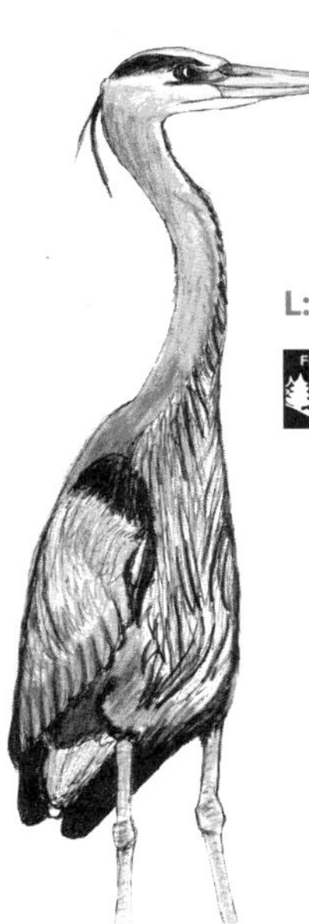

L: 46" W: 72"

American Bittern

Botaurus lentiginosus

Streaked tan and white to blend in with the emergent plants of their watery environment. It's easy to unknowingly approach the stock-still bittern and then be surprised as it explodes into the air with a booming bullfrog-like call. Standing with its bill nearly vertical, the bird has a 360-degree view of potential prey and predators, so it can react quickly.

L: 28" W: 42"

Herons, egrets, and bitterns stand quietly or wade slowly until they have the opportunity to stab or snatch a passing fish, frog, or crayfish. Some species spread their wings to attract aquatic prey to the shade or drop pieces of vegetation to bait small fish. Fishing is aided by their polarized eyesight, which reduces the glare of sunlight.

Tundra Swan

Cygnus columbianus

L: 52" W: 66"

These large, long-necked white birds have a black bill with an inconspicuous yellow spot at the base. Arctic nesters, they winter along the Pacific Coast from Vancouver to northern California, from the ocean inland. Their deep, resonant voice comes from a coiled trachea at the base of the neck. On their breeding grounds they tend to be dabblers, eating invertebrates and leaves of aquatic vegetation. On their wintering grounds they are mostly grazers, eating corn, rice, soybeans, or other crops left over after fall harvesting.

In North America the Tundra Swan is sometimes referred to as the Whistling Swan; in Europe it is sometimes referred to as Bewick's Swan. But it's the same species in both locales.

Canada and Cackling Goose

Branta canadensis and *B. hutchinsii*

Found throughout the US and common in the Central Valley of California in the winter. Wrongly called Canadian Geese (because all geese nest in Canada), the Canada Goose has become two species: the larger Canada Goose and the much smaller Cackling Goose. There are also several races of the Canada Goose, such as Dusky, Lesser, and Aleutian that differ slightly in size and coloration. They graze on grass, grains, and aquatic invertebrates. Lays 5–6 eggs in nest on ground, on a natural or artificial platform, and even in trees. Identified by its gray body, black neck, white throat patch, and raspy honk.

Cackling Goose

L: 25" W: 42"

Wood Duck

Aix sponsa

One of the most attractive ducks, the male has a striking multicolored head, red eyes, flowing green nape feathers edged by white, and a white neck and chin. The female, like most female ducks, is a dull brown. Unlike most ducks, wood ducks nest in tree cavities or artificial nest boxes. Adults fly into the nest without perching. The fluffy precocial young jump from the tree, sometimes dozens of feet high, and tumble to the ground. They feed on vegetation on the surface of the water or slightly below by dabbling.

L: 19" W: 30"

Mallard

Anas platyrhynchos

The Mallard is probably the most familiar of all ducks, with the male's green head, white neck ring, cinnamon chest, and curly tail feathers; the female is streaked brown. Its familiar quack is the classic duck sound. It feeds by tipping its head down while paddling its feet, a behavior known as dabbling. Other ducks dive. They feed on aquatic vegetation or invertebrates. The female makes a feather-lined nest in grass and lays 8–15 eggs, which she incubates while the male is elsewhere. They hatch about a month later; the female provides parental care.

Northern Pintail

Anas acuta

L: 21" W: 34"

Marshes

The pintail gets its name from the male's long, pointed tail. He also has a white chest with a stripe of white running up the neck. The female is a mottled brown. Prized by hunters who call it "sprig" after the curly tail, which resembles a plant sprig.

Most ducks can launch themselves almost vertically out of the water or off the land because they have a set of special feathers on their "thumb" called an alula. This alula acts like the slats on the front of a jet plane, smoothing airflow over the wing to prevent stalling. Remove the alula and the bird is grounded!

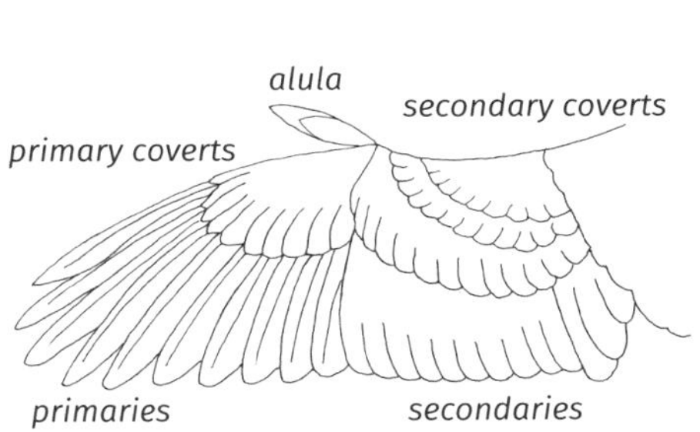

Turkey Vulture

Cathartes aura

The naked red head makes it look smaller than a similar-size hawk. All black, but rear part of wings looks gray while soaring. Differs from hawk and eagle in flight in that the vulture's wings are more V-shaped and the tips of the wings bend upward (a dihedral shape). A decomposer, the vulture cleans the environment by disposing of the bodies of deceased squirrels, skunks, raccoons, and such. A good sense of smell enables it to find food. A very acidic stomach and a habit of defecating on its own feet (to cool off) keep the bacterial load down. Wards off attackers by vomiting. Migration is triggered by change in day length. Hinckley, Ohio, celebrates the annual return of Turkey Vultures on Buzzard Day in February.

L: 26" W: 67"

Red-tailed Hawk

Buteo jamaicensis

A common hawk found throughout North America, the adult has a rusty-colored, not red, tail. Immature birds have a banded tail until 3 years old. Body darkness varies from very light to very dark, but all Red-tailed Hawks show a darker breast band over a lighter chest. Often seen soaring over open fields or perched on fences or utility poles. Feeds mainly on rodents, such as rats and squirrels, but will also eat lizards, frogs, and other birds. Builds large, open nests. Once called the chicken hawk, the Red-tailed Hawk's ecological role in pest management outweighs the minimal damage done to chicken coops. The Red-tailed Hawk is protected by federal law.

L: 19" W: 49"

Bald Eagle

Haliaeetus leucocephalus

The Bald Eagle, our nation's symbol, was once endangered by DDT but is now recovered. An opportunistic feeder, the bald eagle eats a diet of more than 50% fish. Most of the rest is mammals and birds, but it may scavenge food scraps or carcasses. Found all over North America, it is common on garbage dumps in vulture-free Alaska. It builds large nests that it reuses each year, often resulting in a nest 10 feet wide and weighing 2 tons. The white head and tail come at 4–5 years of age; the "bald" name comes from "piebald," referring to the patches of white.

L: 31" W: 80"

Most hawks and eagles lay three eggs, which hatch in order of their laying. This results in different-sized chicks that compete for food from the parents. In times of food scarcity, only the older, larger chicks survive.

American Kestrel

Falco sparverius

The smallest American falcon is found in much of North and South America. Males have blue-gray wings with black spots and white undersides with black barring. The back is rufous and barred on the lower half. The tail is rufous, with a white or rufous tip and a black band. The face has two narrow, vertical black facial markings on each side. Feeds on small animals such as grasshoppers, crickets, lizards, mice, and small birds. It is often seen along roadsides and fences hovering with quick wingbeats as it scans for prey. The kestrel attacks most of its prey on the ground, except for birds, which it captures in midflight.

L: 10" W: 22"

Prairies

Woods

Farms, Parks, Cities

California Quail

Callipepla californica

The California state bird is distinguished by a dark-black chin, gray breast, and six plumes on its forehead; females are brownish with smaller plumes. Prefers dense vegetation. Call sounds like "Chi-ca-go, Chi-ca-go." Does not need to drink because water in food—seeds, grain, berries—is sufficient. A sociable bird, it is often found in groups called coveys, which take daily communal dust baths to absorb excess body oil from their feathers after preening. The similar desert-dwelling **Gambel's Quail** (*Callipepla gambelii*; L: 10", W: 14") has a black spot on its lower belly, and the **Mountain Quail** (*Oreortyx pictus*; L: 11", W: 16") has a very tall head plume; found at higher elevations. All are fast runners and quick flyers.

L: 10" **W: 14"**

Woods Prairies

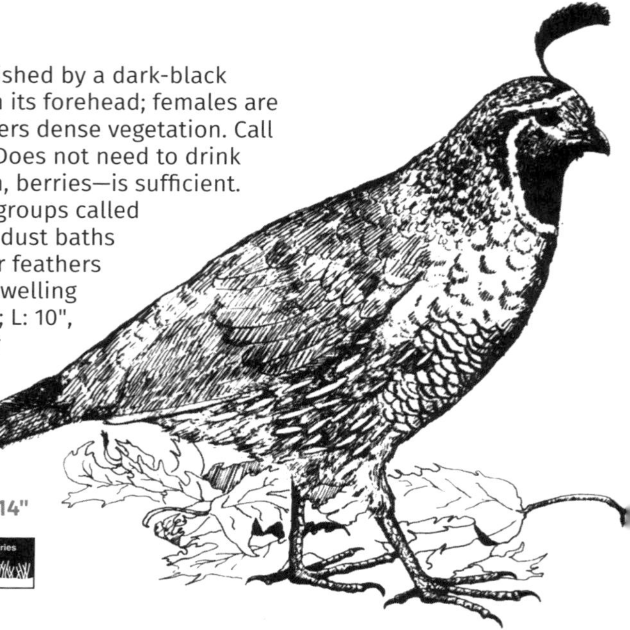

Wild Turkey

Meleagris gallopavo

The Wild Turkey inhabits open woodlands. Adults have long legs and dark feathers with a coppery sheen. Adult males have a large, naked, reddish head; red throat; red wattles on the head and neck; and a long fan-shaped tail. The duller female is much smaller. They are omnivorous, foraging on the ground or in small shrubs for acorns, nuts, and berries. Generally found in groups with one or two males and several females and young. Ground dwelling, they can fly quickly when startled.

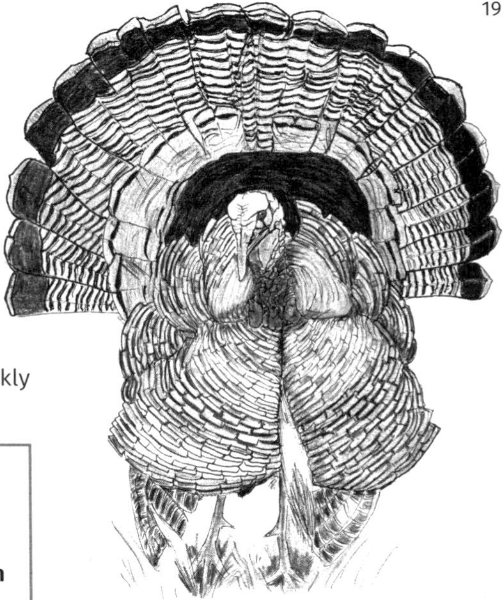

Quail, pheasants, turkeys, and chickens can digest hard grains and nuts. An expandable crop in the throat provides some predigestion, then the food moves to the glandular part of the stomach for more digestion, then into the muscular gizzard where it is ground by gravel that has been swallowed (grit).

L: 46"   W: 64"

American Coot

Fulica americana

A common denizen of watery areas, the all-black coot has white wing edges and a white bill with a black ring. Long, lobed toes help it swim. Stubby wings require long runs on the water before takeoff, but often the runs are sufficient. Often called white-bills or mud hens, they lay 6–7 eggs on a mat of floating vegetation. Young swim soon after hatching and feed on the coot's usual diet of aquatic organisms and vegetation. Vaguely resembling a duck, coots are a member of the rail family and a distant cousin of cranes. A hunted bird, it is easy to shoot but not good to eat.

L: 15" W: 24"

Ocean | Freshwater | Marshes

Killdeer

Charadrius vociferus

The Killdeer's common name comes from its loud call, “kill-deer, kill-deer.” Upperparts are mostly brown, head has patches of white and black, and the white breast has two black bands. Rufous tail and rump. A common shorebird found in grassy and open habitats, including grasslands, schoolyards, and golf courses. Eats various invertebrates. Simple nest is just a depression in gravel. Four spotted eggs are positioned with ends inward to prevent rolling; if one egg is destroyed, parents replace it with a rock. Young are speckled like eggs and can run quickly, although will freeze if approached by an intruder. To fool predators, parents will often feign injury by drooping a wing and leading the intruder away.

American Avocet

Recurvirostra americana

The scientific name describes the bird's most obvious feature, the black and pointed curved bill. Its black-and-white wings and rufous head and neck are good identifying marks. The American Avocet tends to prefer habitats with fine sediments for foraging. Made for probing deep in the mud and skimming invertebrates off the surface of the water, the bird will also just peck at small prey. Although a long-legged shorebird, it will also swim in deeper water to forage. The similarly shaped **Black-necked Stilt** (*Himantopus mexicanus*; L: 14", W: 29") has a shorter, straighter bill; black upperparts and head; white underparts; and white surrounding the eye.

L: 15" W: 31"

Marshes

Spotted Sandpiper

Actitis macularius

L: 7" W: 15"

Ocean Freshwater

The most common sandpiper in North America, the Spotted Sandpiper is also one of the most common sandpipers along rivers and streams. Its short tail, bobbing behavior, and white breast with brown spots give it away. Rarely walks but runs in short spurts and flies, alternating quick flapping and gliding. Lays four spotted, well-camouflaged eggs in a simple depression in gravel. There are many other sandpipers and shorebirds along the Pacific Coast. Some are difficult to identify, with small ones lumped into a nondescript category of "peeps."

Head or body bobbing is common among some birds, and it is thought to give the bird a better perspective of its environment, much as a golfer looks at the green to line up a putt. It can also be part of courtship behavior.

Herring Gull

Larus argentatus

One of the most common gulls in North America, the bird has a light-gray back; peach-colored legs; and a short, hooked bill. Often sitting on water, they manage to keep their inner feathers dry. Gulls eat almost anything, dead or alive, and often become pests at landfills and picnics. Build nests on rocky lakeshores, islands, or cliff ledges. Two or three young hatch and peck at the red spot on the parent's bill to stimulate food regurgitation.

Other Pacific Coast Gulls

Western (*L. occidentalis*; L: 25", W: 58") dark back, peach-colored legs

California (*L. californicus*; L: 21", W: 54") light back, yellow-green legs

Ring-billed (*L. delawarensis*; L: 17", W: 48") black ring around yellow bill

Bonaparte's (*Chroicocephalus Philadelphia*; L: 13", W: 33") black head and fast, tern-like flight

“Seagulls” may range far inland.

Black Tern
Chlidonias niger

Terns are related to gulls but are smaller, with longer and thinner wings and a straight, pointed bill. The only all-black tern on the Pacific Coast. Graceful flight over water becomes erratic due to frequent turning to look for fishy prey. Do not sit on water surface. May hover before diving. Aggressive at nest site of floating reeds and will dive-bomb intruders. North American Black Terns migrate to the coasts of northern South America.

L: 10" W: 24"

Freshwater

Other commonly seen terns are **Forster's** (*Sterna forsteri*; L: 13", W: 31") and **Caspian** (*Hydroprogne caspia*; L: 21", W: 50"). Both are white with a black cap. Caspian is twice as large with a blood-red bill.

Mourning Dove

Zenaida macroura

The Mourning, not morning, Dove is so named for its plaintive call: a mournful hooting. Mourning Doves are grayish brown with black splotches on their wings and black markings on their head. With a pointed tail, it is smaller and faster in flight than the Rock Pigeon with its nearly squared-off tail. Often sits on wires and fences. Like other pigeon relatives, it lays two eggs in a saucer-shaped nest on or near the ground. The naked, helpless young are fed "pigeon milk," the sloughed-off lining of the parent's crop. Common across the continent, Mourning Doves may gather in large flocks during migration.

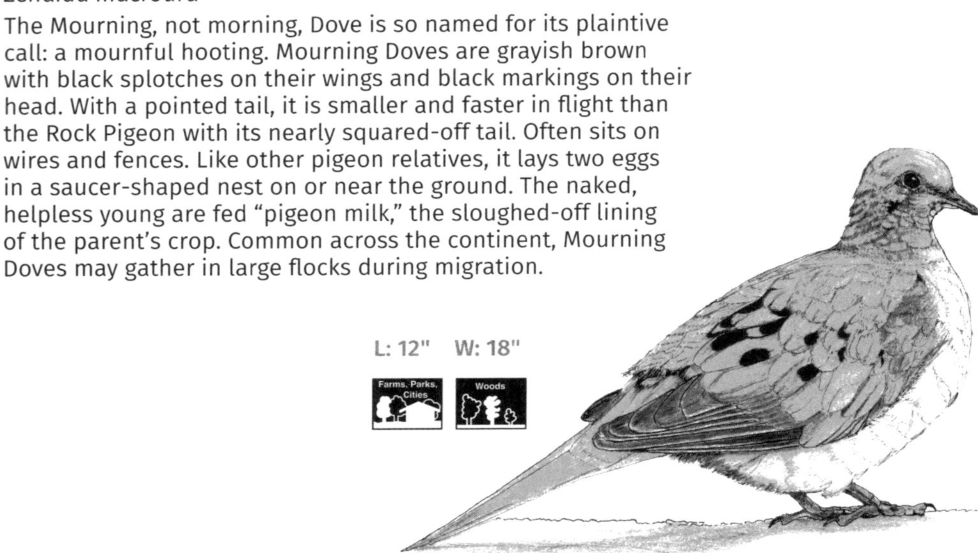

L: 12"   W: 18"

Eurasian Collared-Dove

Streptopelia decaocto

Native to Asia and Europe, this dove was introduced into Japan and North America, where it spread quickly. Having six or more broods a year, it has become quite common and, in some states, can be hunted year-round with no limit. It apparently has had no negative effect on native birds so far. A light-gray color in the front, light tan on the back, and a black half ring on the back of the neck identify it. Its "coo-COO-cook" call is distinctive.

L: 13" W: 22"

Farms, Parks, Cities

There is no biological difference between pigeons and doves, although generally pigeons are larger. Pigeons and doves, unlike other birds, can drink water by sucking with their head down.

Rock Pigeon

Columba livia

Found all over the US, the Rock Pigeon, once called Rock Dove, is the common pigeon found around the world. Color varies from white to black, with a mottled-gray color being most common. Has an iridescent green neck and a structure over the base of the bill called an operculum. A city bird, it eats most anything, although it prefers grains. Nesting on rooftops and window ledges that remind them of their original cliff-dwelling habitat in Europe and Asia, they might have six broods. Since ancient Egypt, pigeons have been bred into over 1,000 varieties.

Pigeons, starlings, and house sparrows are highly adapted to cities and farms, thrive on a variety of food, proliferate for a lack of natural predators, make messy nests, and are difficult to control.

L: 12" W: 28"

Barn Owl

Tyto alba

One of the most widely distributed and common birds in the world, this medium-size, pale tan owl has long wings, a short square tail, and a heart-shaped face. The head and upper body typically vary between pale brown and some shade of gray. Searches at night over prairies and farms for rodents, using night vision that far exceeds that of humans. Humanlike ears, hidden beneath feathers, can accurately pinpoint the source of the quietest sound. Fringed flight feathers make for quiet flight. Like hawks and other owls, eggs hatch in the order they are laid, resulting in different-aged young that compete for food in the nest. Owls regurgitate indigestible fur, feathers, and bones, producing owl pellets; look for them under the nesting tree.

L: 16" W: 42"

Woods | Farms, Parks, Cities

Anna's Hummingbird

Calypte anna

L: 4" W: 5.25"

Farms, Parks, Cities Woods

Anna's Hummingbird is common along the Pacific Coast. The male's head and throat are an iridescent red. They can hover and even fly backward with strong muscles and wings, beating at 60 wingbeats per second. Primarily eat flower nectar obtained by extending a tube-shaped tongue with mop-like fringes and squeezing out the nectar as the tongue is withdrawn. Aggressively defends food source. Use a feeder with a 4:1 sugar solution to attract them. During breeding season, insects provide protein for the growing young. The tiny cup-shaped nest of fibers holds two eggs (it can be covered with a quarter).

Other common hummingbirds

Allen's (*Selasphorus sasin*; L: 3.75", W: 4.25") has rusty feathers with a green back and pink around the chin.

Rufous (*Selasphorus rufus*; L: 3.75", W: 4.5") is almost identical to Allen's but with broader tailfeathers.

Black-chinned (*Archilocus alexandri*; L: 3.75", W: 4.75") is distinguished by a white chest, purple neck, and black chin; females are difficult to identify.

Belted Kingfisher

Megaceryle alcyon

Found throughout North America, the shaggy-headed blue-and-white kingfisher is easy to identify, although only the female has the rusty chest belt. Its long, chattering flight call is distinctive. Perches at stream sides with a good view of the water and dives for fish; may even swim a bit. Brings fishy prey back to perch and beats it on a branch before swallowing it headfirst, to avoid fish getting stuck in the throat. Usually alone, except during breeding season when a pair digs a tunnel in a stream bank, assisted by two toes that are fused together. Lays about seven white eggs that need no camouflage because they are far back in the burrow.

L: 13" W: 20"

Freshwater

Acorn Woodpecker

Melanerpes formicivorous

L: 9" W: 17"

A characteristic bird of the Pacific Coast south of Washington, the Acorn Woodpecker is common in oak woodlands and many suburbs. Back of head is red, forehead and throat white, back is black; shows white wing patches and rump in flight. Live in large groups that raise young cooperatively and share "granaries"—trees, wood poles—that they drill holes in to store their acorns and nuts tightly to deter squirrels, jays, and crows. They eventually crack the nuts open to extract the meat. The **Downy** (*Dryobates pubescens*; L: 7", W: 12") and **Hairy Woodpeckers** (*Dryobates villosus*; L: 9", W: 15") are mainly black on upperparts and wings with a white back, throat, and belly. Adult males have a red patch on the back of the head.

Woodpeckers hold onto tree bark vertically using two toes front and two toes back and a long tail with stiffened feathers. They extend their tongues to explore crevices for insects, larvae, and eggs. A special jaw mechanism absorbs the shock of pecking, as do compressible spongelike bones in the skull. Woodpeckers proclaim territorial rights by pounding on hollow trees, wooden or even metal utility poles, and sometimes houses.

Northern Flicker

Colaptes auratus

This common large woodpecker is often seen probing the ground. Brown with black bars on the back and wings, a necklace-like black patch on the upper breast, and beige lower breast and belly with black spots. Males can be identified by a red "mustache" at the base of the beak. The white rump is conspicuous in flight. Because the underside of the wing can vary from red to orange to yellow (it's mainly red on the Pacific Coast), the Northern Flicker was once thought to be two species.

The Northern Flicker is often seen sitting on an ant hill, allowing ants to crawl over its body. The bird then grabs one ant, crushes it, and rubs it through its feathers. It may repeat this behavior several times. The ants contain chemicals, such as formic acid, that can kill insects, mites, fungi, and bacteria. Many other species of birds engage in "anting."

L: 12" W: 20"

Western Kingbird

Tyrannus verticalis

Found throughout the western half of the US in the summer, the Western Kingbird has a gray back, dark wings, white chin, yellow belly, and black tail. Commonly perched along rural roadsides on wires or fences. Diet largely consists of flying insects caught by ambushing them in midair, a feeding behavior called sallying. *Kingbird* refers to their aggressive behavior, defending their territory even against much larger birds. They will attack humans, livestock, and dogs if they think their young are in danger. The kingbird is in the flycatcher family *Tyrannidae*, the largest bird family in the world, characterized by a flattened, triangular beak with a downward hook at the end of the upper bill.

Loggerhead Shrike

Lanius ludovicianus

Resembling a stocky mockingbird with gray, white, and black coloration, the Loggerhead Shrike has a shorter tail and a distinctly hooked beak. A white throat and black stripe through the eye are identifying. Shrikes behave like hawks. Often alone in the open on fences, wires, or the tops of small trees, especially near farms, they scan for prey. With rapid wingbeats, they attack sitting birds, small mammals, and large insects. Their songbird feet are too weak to serve as talons, so they impale their prey on small branches or barbed wire, earning them the colloquial name of butcherbird. *Loggerhead* refers to the disproportionately large head as compared to the body. The **Northern Shrike** (*Lanius excubitor*; L: 10", W: 14.5") is similar but larger and only seen in the winter.

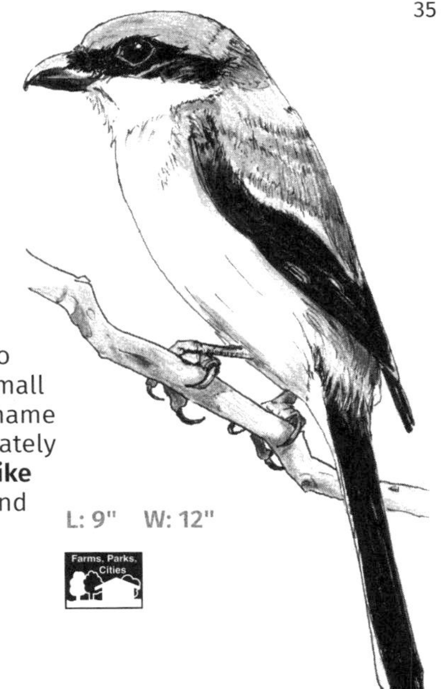

L: 9" W: 12"

Farms, Parks, Cities

Steller's Jay

Cyanocitta stelleri

Blue with a black head with a distinct topknot, the Steller's Jay is common in coniferous forests of higher elevations and lower down in the evergreen forests of the Pacific Coast foothills. Moves to lower regions during cold winters. Invades campgrounds and hops around, hoping to steal food. Generally, it is omnivorous, eating small birds, insects, berries, and other birds' eggs and young. Steller was a German naturalist who worked in Russia; he has several plants and animals named after him.

The blue color of birds does not come from pigments as other colors do. Blue color is created by the way light interacts with the cells of the feathers; blue is a structural color. Look down at a blue feather and it looks blue, but hold it up to the light and you will see it is actually brown.

California Scrub-Jay

Aphelocoma californica

The California Scrub-Jay has a blue head, wings, and tail; gray-brown back; and grayish underparts. It is common in parks, backyards, and birdfeeders at lower elevations. They usually forage in pairs or family groups and eat small animals, as well as insects, nuts, and berries. They will tear open the bottom of another bird's nest and eat the eggs or young, giving the jay an unsavory reputation. The California Scrub-Jay is not a Blue Jay, a bird mainly restricted to the eastern half of the US.

L: 11" W: 15"

Farms, Parks, Cities

Woods

American Crow

Corvus brachyrhynchos

Black with a rounded tail in flight, the crow's nasally, high-pitched call is a loud "caw." Not very pleasant, but crows are songbirds with a complex voice box and have many different calls. Migrates. Winter flocks may number in the millions. Inhabits virtually all types of country. Omnivorous, eating almost anything it can get in its beak—insects, small birds and mammals, worms, nuts, berries, and even carrion. Their black feathers are actually iridescent and seem to change colors as light hits them from different angles.

L: 17" W: 39"

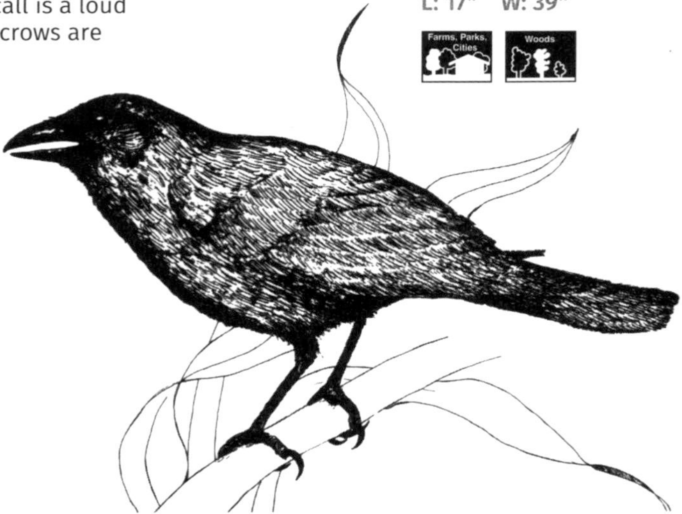

Farms, Parks, Cities

Woods

Common Raven

Corvus corax

About a third larger than the American crow, the raven also has a heavier head, a longer bill, a tuft of bristles over the bill, and is perhaps the largest songbird at 2.5 pounds. Also omnivorous like the crow, the raven prefers wooded areas with open habitat nearby. It has been expanding its habitat to increasingly overlap that of the crow. In flight, the raven has a wedge-shaped tail. The raven has been revered in many cultures as a spiritual symbol. It has a low, hoarse call. Will share large carcasses with wolves.

L: 24" W: 53"

Barn Swallow

Hirundo rustica

Found over most of the world, it is the only swallow in North America with a long, forked tail. Metallic blue above with a cinnamon throat and rusty belly, it resembles the **Cliff Swallow** (*Petrochelidon pyrrhonota*; L: 5", W: 13"), which lacks the long tail. Nests in colonies up to several hundred birds, placing grass- or feather-lined nests made of mud and straw in, on, or under bridges or farm buildings. Like all swallows, it feeds on insects while in flight.

Cliff Swallows make elaborate mud nests; **Bank** (*Riparia riparia*; L: 5", W: 13") and **Northern Rough-winged Swallows** (*Stelgidopteryx serripennis*; L: 5", W: 14") use riverbank burrows; and **Tree** (*Tachycineta bicolor*; L: 5", W: 14") and **Violet-green Swallows** (*Tachycineta thallasina*; L: 5", W: 13") nest in tree cavities.

Swallows spend the summer near water and relocate to the southern continents for the winter.

Oak and Juniper Titmouse

Baeolophus inornatus and *Baeolophus ridgwayi*

L: 6" W: 9"

Woods

The name is from the Old English *tit* and *mase*, meaning small bird. Two species split from the original plain titmouse, they differ slightly in coloration and call. The Oak Titmouse is common throughout most of California and southern Oregon, while the Juniper Titmouse is found east of the Sierras into the Great Plains. Both are drab gray with a short conical bill and topknot. The Oak Titmouse voices variable calls and songs, often sounding like "peter, peter, peter," while the voice of the Juniper Titmouse is lower and faster. Voracious eaters, they digest a quarter of their weight every day in insects, seeds, and berries. Lay 6–7 eggs in a feather- or hair-lined nest in a tree cavity.

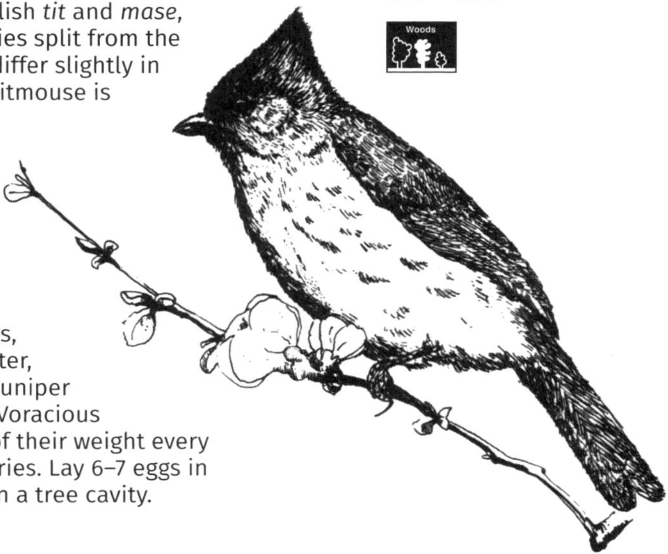

Bushtit

Psaltriparus minimus

Residents of shrubby or oak woodland, Bushtits are often found during the winter in lively groups of 6–10 birds, flitting from branch to branch, often hanging upside down, looking for invertebrates. They have grayish-brown rotund bodies with brown heads and longish tails. In spring, a pair will build a long, hanging, woven nest with an entrance on the side near the bottom. The swaying of the nests, which hold 5–10 eggs, prevents the egg membranes from sticking to the shell. Glue from the parents' salivary glands ensures the eggs won't break from knocking into each other. Females from other nests help each other incubate their eggs.

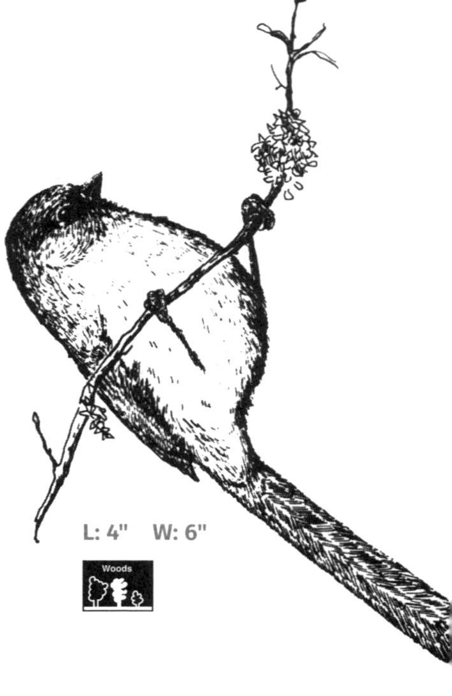

White-breasted Nuthatch

Sitta carolinensis

A deciduous forest resident, it has a gray body; white belly, breast, and face; a black head and nape; and a chisel-like beak. Most often seen spiraling down a tree trunk, one foot at a time, probing the bark for invertebrates, seeds, and nuts. It has a large back toe that helps it cling to the bark. Reaching the bottom, it flies to the next tree and repeats. Its name comes from the habit of jamming a nut into tree bark and then "hatching" out the seed by pecking at it. Mainly coniferous forest inhabitants, the similar **Red-breasted Nuthatch** (*Sitta canadensis*; L: 4", W: 8") is smaller, with a white stripe over its eyes, a black stripe through its eyes, and reddish undersides. The **Pygmy Nuthatch** (*Sitta pygmaea*; L: 4", W: 8") is grayish with a brown cap and black eye stripe.

L: 6" W: 11"

House Wren

Troglodytes aedon

Found in woods and cities, this small grayish-brown bird with black barring on the wings and tail has a long, thin, downcurved bill for insect eating. Sings long, complex, melodious songs like all wrens. Often sings from exposed perches with tail cocked upright. Nests in tree cavities, rock walls, and emergent vegetation. **Bewick's Wren** (*Thyromanes bewickii*; L: 5", W: 7") is brownish on top, grayish underneath, and has a white stripe over the eye. Common in wetlands, the **Marsh Wren** (Cis*tothorus palustris*; L: 5", W: 6") has brown upperparts, a light-brown belly and flanks, a white throat and breast, a black back with white stripes, a dark cap, and a white line over the eyes. Wrens tend to be polygamous in times of abundant food. Females raise young unassisted.

L: 5" W: 6"

Woods | Farms, Parks, Cities | Marshes

American Dipper

Cinclus mexicanus

L: 7" W: 11"

Freshwater

The stocky, gray-bodied American Dipper lives from the bottom of California to the top of Alaska in fast-moving mountain streams or cold coastal ones. Once called the Water Ouzel. Swims under water, probing rocks for larvae. Streamlined wings, thick insulation, an extra eyelid, scales that close over the nostrils, and a large oil gland to preen and waterproof the feathers allow this songbird to submerge for 20 seconds. Rarely flies. Lays 3–5 eggs in a nearly spherical nest at water's edge or even behind a waterfall. Complex song is flutelike.

Ruby-crowned Kinglet

Corthylio calendula

This small bird is common in many areas of the US. Often seen actively flitting from branch to branch in search of insects, the male will occasionally raise his head feathers to expose a ruby-red crown when agitated or displaying. Gray-green above and lighter underneath, the bird's white wing bars and eye ring are distinctive. Immature males have a yellow crown; the female has none. They nest in a hidden hanging cup suspended from a tree branch and may have up to 12 eggs in a clutch, a very large number for such a small bird.

Western Bluebird

Sialia Mexicana

The head, wings, and back are blue; the chest and abdomen are orange, except for a small whitish belly. Nests in tree cavities in low- and mid-level woods and nest boxes. Eats insects and berries and helps distribute mistletoe. Forms winter flocks, sometimes with the **Mountain Bluebird** (*S. corrucoides*; L: 7", W: 13"), which nests at high subalpine elevations.

Bluebird males are far more colorful than their female mates, typical of the bird world. Males display to attract females and defend their territory, while the dull-colored females need to avoid attracting predators while they attend to parental duties on the nest.

L: 7" W: 14"

Woods Forests

American Robin

Turdus migratorius

Perhaps the best-known American bird, the American Robin is identified by its black head, dark-gray back, and rufous breast and belly. Known as the harbingers of spring, robins hold winter territories even in very cold environments. They feed on the ground and can hear earthworms moving through their tunnels. Very aggressive during nesting season and may attack its reflection in windows and car mirrors. Lays 3–5 light-blue eggs in a nest made of mud and grass.

Young robins (and other songbirds) hatch in about 10 days. After another 10 days they will jump from the nest, even though they can't fly. Parents will take care of juveniles on the ground until their feathers grow enough to allow them to fly. People mistakenly think baby robins have fallen from their nest and need help. They don't.

L: 10" W: 17"

Woods | Forests | Farms, Parks, Cities

Northern Mockingbird

Mimus polyglottos

The Northern Mockingbird—the state bird of five states—is gray on the back and lighter on the front, with white wing patches in flight. Its long tail is black in the center and white on the sides. Mimics other birds with a varied repertoire, each song repeated once. It will often sing all night, especially during a full moon. An omnivore, it eats mainly insects and fruits. Intelligent, mockingbirds can recognize individual humans. It prefers open areas with little vegetation. Once restricted to the southern states, mockingbirds are now found across the US and even in parts of Canada. Climate change has warmed the environment, so many birds are moving their ranges northward.

L: 10"   W: 14"

European Starling

Sturnus vulgaris

Imported from Asia into New York's Central Park in the late 1800s, these starlings later spread to the Pacific Coast in less than a century, becoming pests in many urban areas. Starlings have a shorter tail than other all-black birds; in fall, they have a black bill and blue-black metallic sheen to the body. In spring, the bill turns yellow and wear on the body feathers exposes numerous white spots. Gives off a series of high whistles and, related to mynahs, is a good mimic. Eats insects and grains. Nests in cavities and often outcompetes native birds for nest holes.

L: 8" W: 16"

Farms, Parks, Cities

The European starling, house sparrow, and rock pigeon are not protected by federal law, as all other birds are, because they are nonnative. The starling was imported by the American Acclimatization Society, which introduced European plants and animals into North America for economic and cultural reasons.

Cedar Waxwing

Bombycilla cedrorum

The Cedar Waxwing, a resident of the northern US where it breeds in coniferous forests, is recognized by its silky plumage, black face mask, and permanent topknot. The tips of the secondary wing feathers appear to have been dipped in sealing wax. The more red the tips a bird has, the more mature it is. Winter visitors in the South, they're most often seen in flocks descending on berry bushes to gorge themselves. Social creatures, a satiated bird will pass a berry onto its neighbor. Occasionally, older fermented *Pyracantha* berries will cause the birds to become tipsy and even fall from their perch. In summer, they prefer insects.

L: 7" W: 12"

SUMMER

WINTER

Yellow-rumped Warbler

Setophaga coronata

The most common warbler in winter. Whether in its striking summer plumage with a black chest or its brownish-gray winter outfit, it can always be identified by the yellow rump, throat, and flanks. Winters in the California Central Valley and along the coast and summers in higher-elevation coniferous forests. Flits rapidly through vegetation, picking insects and larvae from twigs and leaves. They often sally out, flycatcher-like, to capture insects in midflight. Warblers have complex "warbling" songs and thin bills for reaching into bark crevices for small food items.

Most warblers migrate to the southern states or Central or South America to spend the winter and seek invertebrates and fleshy berries. The Yellow-rumped Warbler, however, is able to digest the waxes found in wax myrtles, indigestible by most other birds, allowing it to winter farther north than any other warbler.

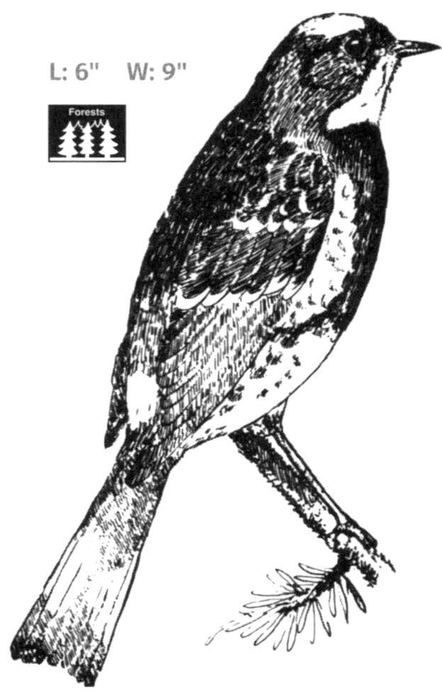

Spotted Towhee

Pipilio maculatus

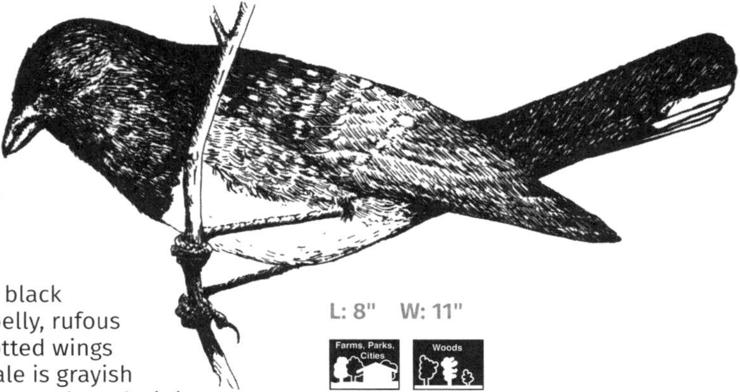

The name towhee is imitative of the call of the **Eastern Towhee** (*Pipilo erythrophthalmus*; L: 8", W: 11"). Often, the only sound heard is a short "mew" given from the underbrush. The male's black head, white chest and belly, rufous sides, red eyes, and spotted wings are distinctive; the female is grayish brown. Feeding on insects and seeds, it is most frequently seen near the ground or in a low bush in open forests, brushy fields, and chaparrals, noisily rummaging through dry leaves, looking for food. It breeds across northwestern North America and is present year-round on the Pacific Coast. The similarly shaped **California Towhee** (*Melozone crissalis*; L: 9", W: 11") is dull brown with light-rust undertail feathers and restricted to that state. Neither towhee is migratory.

L: 8" W: 11"

Farms, Parks, Cities

Woods

White-crowned Sparrow

Zonotrichia leucophrys

A winter visitor, the adult has a gray face and black-and-white–striped head; immatures have brown stripes on a gray head. Breeds in tundra, where it nests on the ground. Common in low-elevation wooded areas, parks, and backyards. Forages on the ground and in low shrubs for seeds and insects; common at birdfeeders. Has plaintive, fluctuating whistle, which varies with geographic locale and human activity. The **Golden-crowned Sparrow** (*Z. atricapilla*; L: 7", W: 10") has a gray body and a yellow cap bordered by black; the immature, resembling a large female House Sparrow, has a yellowish forehead. Often found in mixed flocks with the White-crowned Sparrow.

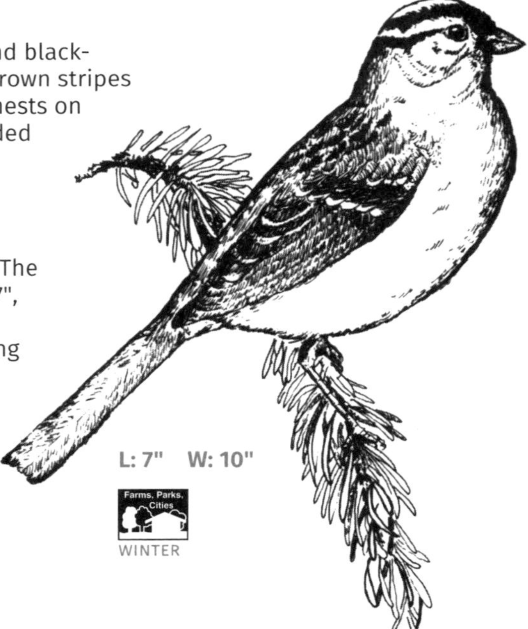

L: 7" W: 10"

Farms, Parks, Cities

WINTER

Dark-eyed Junco

Junco hyemalis

The Dark-eyed Junco is variable in color but basically grayish brown with a pink bill and white belly. The head and neck are darker, often black, resembling an executioner's hood. Breeds in northern coniferous forests and winters on the Pacific Coast. Related to similarly sized sparrows with which it often associates, it hops around on the ground and in brush piles, looking for seeds and insects. In flight, you can see its bright-white outer tail feathers, which disappear upon landing. These white feathers may serve as communication within the flock or as a way of deterring a predator, which loses sight of the junco when the white disappears.

L: 6" W: 9"

WINTER

Western Meadowlark

Sturnella neglecta

A black V on a yellow chest is identifying. Not a real lark but a blackbird, it sings like a lark with a melodious song that helps it attract a mate and declare territory in a treeless grassland. Camouflaged by a brown-streaked back, it weaves its way through the grass, probing for seeds and invertebrates. Five eggs are laid in a domed tunnel of grass. The range of the Western Meadowlark extends to the Midwest, where it overlaps with the **Eastern Meadowlark** (*S. magna*; L: 10", W: 15"); they are virtually identical, except for their voice, which helps to keep the species separate.

L: 10" W: 15"

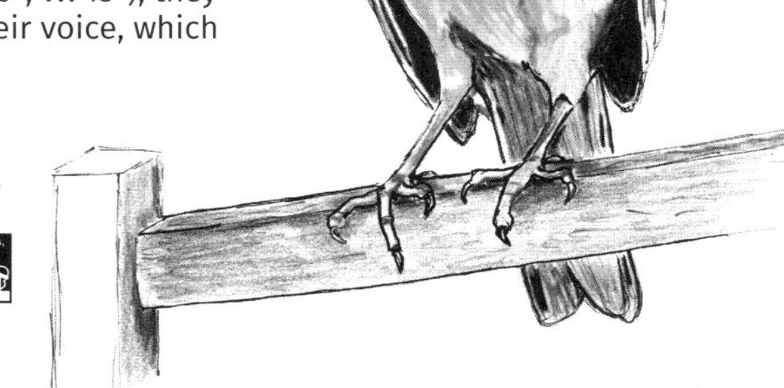

Red-winged Blackbird

Agelaius phoeniceus

One of the most widespread and recognizable of all birds is the "redwing," named for of the ruby-red shoulder patch of the male. Males establish their territories in a wetland by singing and raising the red patch, hoping to attract 1–3 sparrow-looking, streaked-brown females. Experiments in which the red shoulder patch was painted black caused the males to lose their territories to competitors. Each female builds a nest supported over water by emergent vegetation, lays four eggs, and incubates them. Red-winged Blackbirds spend the winter in separate male and female flocks.

L: 9" W: 13"

Marshes

Brewer's Blackbird

Euphagus cyanocephalus

A common bird of fields and farms, the glossy-black male has a bluish-black head and yellow eyes; the brownish female has brown eyes. It forages on the ground for a variety of items, reflected in its genus name *Euphagus*, which means wide diet. *Cyanocephalus*, the species name, means bluish head. Often winters in large flocks along with starlings and other blackbirds. Large flocks, with more eyes, are helpful in avoiding predators and finding food.

Members of the blackbird family poke their bills into the soil and, with strong jaw muscles, push aside the soil, making a hole that allows them access to seeds or invertebrates that they would not find otherwise. This behavior is called gaping.

L: 9"   W: 16"

Farms, Parks, Cities

Evening Grosbeak

Coccothraustes vespertinus

This heavy-bodied bird has a short black tail, black wings, and a large light-colored bill. The adult male has a bright-yellow forehead and body and a large white patch on the wing. The female is a dull gray but has the large white wing patch. The grosbeak nests in coniferous forests but winters in large flocks on the Pacific Coast, seeking large seeds, berries, and insects. Its heavy bill can open pine nuts and olive and cherry pits that are inaccessible to other birds. Its name came from French explorers who thought that it only came out in the evening.

SUMMER WINTER

House Sparrow

Passer domesticus

One of the most familiar of all birds, the House Sparrow, once called the English Sparrow, is found in many urban areas worldwide and is nonmigratory. The male has a black face mask, eye stripe, and nape. The female is dull brown. Introduced into New York in the 1800s, they quickly became pests, and two decades later efforts were made to exterminate them. Today, with the increasing use of pesticides and the decreasing number of insects, the House Sparrow is in decline. Not closely related to other sparrows in the US, it is more akin to the weaver birds of Africa.

L: 6" W: 10"

Farms, Parks, Cities

INDEX

Other books in the pocket-size *Finder* series:

FOR US AND CANADA EAST OF THE ROCKIES

Berry Finder native plants with fleshy fruits
Bird Finder frequently seen birds
Bird Nest Finder aboveground nests
Fern Finder native ferns of the Midwest and Northeast
Flower Finder spring wildflowers and flower families
Life on Intertidal Rocks organisms of the North Atlantic Coast
Scat Finder mammal scat
Track Finder mammal tracks and footprints
Tree Finder native and common introduced trees
Winter Tree Finder leafless winter trees
Winter Weed Finder dry plants in winter

FOR THE PACIFIC COAST

Pacific Coast Fish Finder marine fish of the Pacific Coast
Pacific Coast Mammal Finder mammals, their tracks, skulls, and other signs
Pacific Coast Tree Finder native trees, from Sitka to San Diego

FOR THE PACIFIC COAST *(continued)*

Pacific Intertidal Life organisms of the Pacific Coast
Redwood Region Flower Finder wildflowers of the coastal fog belt of CA

FOR ROCKY MOUNTAIN AND DESERT STATES

Desert Tree Finder desert trees of CA, AZ, and NM
Rocky Mountain Flower Finder wildflowers below tree line
Rocky Mountain Mammal Finder mammals, their tracks, skulls, and other signs
Rocky Mountain Tree Finder native Rocky Mountain trees

FOR STARGAZERS

Constellation Finder patterns in the night sky and star stories

FOR FORAGERS

Mushroom Finder fungi of North America

NATURE STUDY GUIDES are published by AdventureKEEN, 2204 1st Ave. S., Suite 102, Birmingham, AL 35233; 800-678-7006; naturestudy.com. See shop.adventurewithkeen.com for our full line of nature and outdoor activity guides by ADVENTURE PUBLICATIONS, MENASHA RIDGE PRESS, and WILDERNESS PRESS, including many guides for birding, wildflowers, rocks, and trees, plus regional and national parks, hiking, camping, backpacking, and more.